Energy 236

你能变得多快？

How Fast Can You Change?

Gunter Pauli

［比］冈特·鲍利 著

［哥伦］凯瑟琳娜·巴赫 绘

贾龙慧子 译

上海远东出版社

丛书编委会

主　任：贾　峰

副主任：何家振　闫世东　郑立明

委　员：李原原　祝真旭　牛玲娟　梁雅丽　任泽林

　　　　王　岢　陈　卫　郑循如　吴建民　彭　勇

　　　　王梦雨　戴　虹　靳增江　孟　蝶　崔晓晓

特别感谢以下热心人士对童书工作的支持：

匡志强　方　芳　宋小华　解　东　厉　云　李　婧

刘　丹　熊彩虹　罗淑怡　旷　婉　杨　荣　刘学振

何圣霖　王必斗　潘林平　熊志强　廖清州　谭燕宁

王　征　白　纯　张林霞　寿颖慧　罗　佳　傅　俊

胡海朋　白永喆　韦小宏　李　杰　欧　亮

目录

Contents

在古巴海岸线旁的一座小岛上，一只变色龙（一种蜥蜴）刚刚知道他的天敌是另一种比他体型大一倍的蜥蜴，而且和他生存在同一地区。一颗海藻注意到这只变色龙压力很大，便提议道：

"为什么你不离开这片平原呢？去树上生活吧！"

On a small island off the Cuban coast, an anole has just learnt that its main predator, another type of lizard twice its size, is living in the same area. An algae notices how stressed the anole is and suggests:

"Why don't you just leave the plains? Go and live in the trees!"

海藻注意到这只变色龙压力很大……

Algae notices how stressed the anole is ..

以后我不吃昆虫改吃植物吗？

start eating plants instead of insects?

"你是说我得离开这片土地？这里有我最喜欢的食物，美味的蜘蛛。以后我不吃昆虫改吃植物吗？"

"有什么就吃什么，懂得变通才是生存的关键。"

"可是我的爪子不适合爬树。你瞧，在地面上奔跑捕食需要修长的四肢，而对于爬树来说短小的四肢更有优势。"

"You mean I'll have to move away from my favourite food on the ground, all those delicious spiders? And start eating plants instead of insects?"

"Food is food, and being as flexible as one can be is key in life."

"Oh, but my claws are not made for climbing trees. You know, running after one's food on the ground one needs long limbs. For climbing trees, short limbs work better."

"你会很快适应的。别忘了，这是个飓风频发的国家。在风暴中你需要用爪子紧紧抓住树干和枝条。就是紧紧抓住树皮，明白吗？"

"同意，坚持对于生存至关重要。不过，要与这么多同类住在同一棵树上，这树马上就要变得过度拥挤了。"

"没关系，不同类型的变色龙都会为自己找一个独特的栖息地。"

"You will adapt soon enough. Keep in mind that this is hurricane country. You'll need claws to hold onto tree trunks and branches during a storm. To get a firm grip on the bark, you know?"

"Agreed, holding on is a matter of survival. But with so many of us having to live in a tree, it will soon become overcrowded."

"Don't worry, each family of your species will find a special niche for itself."

风暴中你需要用爪子紧紧抓住树干……

Claws to hold onto tree trunks during storms ...

在树的不同部位找到安家的地方……

Found a home range in every part of the tree …

"有些变色龙喜欢爬到树冠上去，有些喜欢在树冠的枝条上；有些可以生活在树干上，或者树木上半部分的树枝上；有的甚至可以在靠近地面的树干上繁衍生息。"

"是的。几代之后你们可以在树的不同部位找到安家的地方，而你们的爪子也会进化以满足你们的需求。"

"Some could make their way up into the canopy, others to twigs in the crown. Some could live on the trunk, or on the branches halfway up the tree. And a few could even populate the trunk closer to the ground."

"Yes, within just a few generations, all will have found a home range in every part of the tree. And your claws will have adjusted to exactly what is needed."

"我估计几代之后我们的身体就能适应在树上生活了。我也能增大我的垂皮了。"

"垂皮？那是什么？"

"就是我的双下巴。人们可能会觉得它长得奇怪，但是鼓起垂皮的确可以帮我们威慑入侵者，因为这能让我们看起来比实际大……可能也更容易吸引雌性的注意。"

"I guess that within a few generations we would all have adapted our bodies accordingly. And I can enlarge my dewlap...."

"A dewlap? What is that?"

"That's my double chin. People may think it looks strange, but extending it does help to deter any intruders, as it makes us look larger than we are ... and perhaps more beautiful in the eyes of the ladies."

让我们看起来比实际大......

Makes us look larger than we are ...

地面上的蜘蛛就会越来越多。

The ground will be covered in spiders.

"好吧，不管你有没有这个双下巴，只要你们都去树上栖息了，地面上的蜘蛛就会越来越多，他们会捉走所有的昆虫。"

"是呀，以后岛上的日子就和现在大不相同了。"

"听着，我们都需要去适应这个新的现实。海藻、变色龙和昆虫都只要短短几代就能快速适应环境，不像人类……"

"Well, with or without double chins, once you are all living in trees, the ground will be covered in spiders – and they will catch all the insects…"

"Yes, life will never be the same on this island."

"Look, we all need to adjust to this new reality. We algae, anoles and insects are able to adapt quickly, within just a few generations – unlike people…"

"正是如此！我们别无选择，只能做好准备来迎接更好时代的到来。快速的变化是很正常的。"

　　"我们藻类对这件事太熟悉了。我们无时无刻不在经历繁荣和萧条的循环。我们藻类是轮虫最喜欢的食物；当我们数量众多的时候，轮虫也会繁育更多的后代。"

"Exactly! We have no choice but to get ready for better times. Rapid change is the standard."

"We algae know that all too well. We are going through cycles of boom and bust all the time. When the rotifers' favourite food – us – is abundant, they have many more babies."

繁荣和萧条的循环

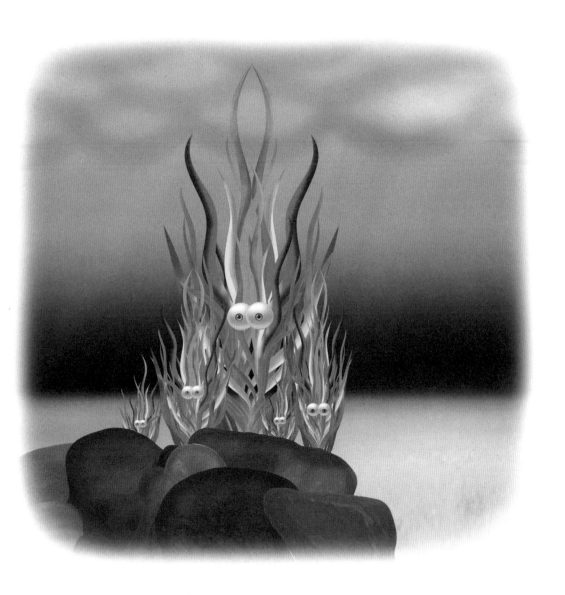

Cycles of boom and bust

……很多饥肠辘辘的轮虫……

... many hungry rotifers ...

"我能想象到那种场面：当有很多饥肠辘辘的轮虫的时候，藻类基本就所剩无几了。而当食物被吃光之后，轮虫也就灭绝了。"

　　"当轮虫几近消亡之后，我们就会重新生长——我们会进化出更多的种类，包括一部分能够抵御轮虫的藻类。"

"I can imagine that there are then so many hungry rotifers that in no time there will be no algae left. The rotifers then die out because they've run out of food."

"After they have nearly vanished, we regrow – but with new algae varieties, including a few that are resistant to rotifers."

"这样一来你们就能从他们的攻击中幸存下来了！太妙了！我们都能改变自身来适应新的环境，环境在变化的同时也塑造了我们。生物进化真是充满惊喜啊！"

……这仅仅是开始！……

"And in that way you survive their onslaught! Brilliant! We can all change to suit our new environment. And when the environment changes, it shapes us. Evolution is just so full of surprises! "

… AND IT HAS ONLY JUST BEGUN!...

⋯ AND IT HAS ONLY JUST BEGUN! ⋯

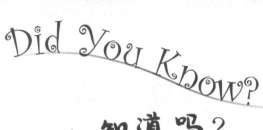

Did You Know?

你知道吗？

Cuba is the main island, with dozens of smaller islands like Cayo Largo, Cayo Coco, Cayo Santa Maria, Cayo Guillermo, Cayo Saetia, and Cayo Levisa. All are world famous for their beautiful beaches.

古巴是一座主岛，拥有许多小岛屿，包括拉戈岛、科科岛、圣玛丽亚岛、尔莫岛、塞蒂亚岛和莱维萨岛等。它们都以靓丽的海滩闻名世界。

The most abundant macroscopic animals in the world may well be the springtail, with an estimated 100,000 individuals found per square meter of soil.

在世界上所有肉眼所能观察到的生物中，丰度最大的可能就属弹尾虫了——每平方米土壤中能够找到大约10万个个体。

22

变色龙是进化的绝佳案例。这些生活在孤岛上的蜥蜴会占据它们觅食的植被上的各类生态位，生活在树木的树冠、树枝或者树干上。

Anole lizards are a great example of evolution. These lizards that live on isolated islands, occupy separate niches within the vegetation where they forage, living in the crown, branches or on the trunk of trees.

每一种变色龙都会发展出自己独有的体型，这被称为生态型。不同的体型能够适应它们所生活的树木的特定部位。树上所有的生态位都被不同种类的蜥蜴所占据。

Each anole lizard family develops its own distinct body type, called an ecomorph, which is adapted to the specific part of the tree in which it lives. All niches of the trees are occupied by different lizard species.

The time-dependent rate phenomenon indicates that the speed of an organism's evolution will depend on the time frame over which the observer is looking at it.

随时间变化的速率现象表明，生物进化的速度将取决于观察者观察它的时间范围。

Humans are now 5-8 centimetres taller than they were 200 years ago, but this doesn't mean that the process will continue, and that we will be a metre taller in 2,000 years.

人类现在比200年前高了5-8厘米，但这并不意味着这个过程会继续下去，即2 000年后，我们会比现在高1米。

Rapid evolution over short periods of just a few generations, does not always persist, and is limited to small populations. These changes can be eliminated or reversed.

在短短几代的时间内迅速进化的这种现象并不总是存在，而且仅限于小种群。这些变化可以被消除或逆转。

In natural selection, evolutionary adaptations are caused by factors such as environmental change, predation or anthropogenic disturbance. These forces must be widespread and continuous to persist.

在自然选择中，进化适应是由环境变化、捕食或人为干扰等因素引起的。这些外力必须是广泛且持续的。

If you could, would you adapt your body to suit the place you're living in? Or do you adapt the house you live in?

如果可以的话，你会改变你的身体以适应你居住的地方吗？或者改变你居住的房子？

How easy would it be for you to move away from your favourite food?

对你来说，离开你最喜欢的食物容易吗？

When you do not understand something, do you ask questions or simply carry on with what you are doing?

当你不理解某些事情的时候，你会选择停下来问问题还是继续你手上的事情？

Is a dewlap something beautiful – in your eyes?

你觉得双下巴漂亮吗？

Are you able to quickly adjust to new realities? When something changes, are you prepared to accept the changes, or are you resistant to change? Ask your parents to tell you more about any drastic changes that your family has had to go through. Also find out whether they feel that, in the end, they came out stronger because of it. Now share those stories with your friends, and listen to their stories, and those of their parents. What do you stand to learn from this exercise?

你能快速适应新的环境吗？当事情发生变化，你会准备接受还是拒绝改变？问问你的父母，看他们有没有经历过一些必须克服的重大改变？经历这些之后有没有变得更加强大？现在和你的朋友一起分享这些故事，也听听他们和他们父母的故事。你从这个经历中能学到什么？

学科知识
Academic Knowledge

生物学	全世界有6 000种蜥蜴；蜥蜴的体形从几厘米到3米不等；垂皮，也叫双下巴；轮虫，也叫轮形动物。
化 学	蜥蜴可以产生9种毒素，蛇也可以；蜥蜴通过信息素来辨认个体；垂皮中的色素是蝶呤和类胡萝卜素。
物 理	蜥蜴拥有视觉、触觉、嗅觉和听觉；蜥蜴是变温动物，拥有一定的调节体温的能力；有蓝色垂皮的蜥蜴可以通过紫外线交流。
工程学	盛衰周期是从短暂的"有害"阶段向更持久的"无害"阶段的发展。
经济学	繁荣与萧条的循环：多年的经济扩张之后，又会有数年的萎缩。
伦理学	我们往往会试图改变周围的一切，但却不喜欢改变自身；有些物种只有在所有个体都能保证食物来源的时候才会繁殖。
历 史	蜥蜴在三叠纪的卡尼期出现，距今大约2.2亿年；沧龙是由水生蜥蜴进化而来的。
地 理	古巴由4 000个岛屿和珊瑚礁组成；古巴90%的爬行动物是蜥蜴。
数 学	脉冲微分方程形式；盛衰周期——人口上升至爆炸水平，然后经历急剧下降。
生活方式	"自然平衡"范例：在生物入侵发生后，我们相信自然会倾向于回归其原始状态。
社会学	蜥蜴在全世界的神话传说和民间故事中都有出现；夏威夷和波利尼西亚群岛有叫莫欧的蜥蜴王；在非洲传说中，乌库鲁库鲁让一只变色龙告诉人类他们可以永生；社会适应性是生存能力的一部分。
心理学	缺乏信心；寻找借口逃避某事；面临危机时为更好的时机做准备的心态；离开家园并居住在新环境中的能力。
系统论	适应环境变化并找到新生态位的能力；随环境改变的能力一开始是暂时的，如果改变足够持久，新的特性就会被编入DNA；如果不够持久，它们会再度改变。

情感智慧
Emotional Intelligence

海藻

海藻对变色龙充满同情。她为他提供了一些建议并做了详细的说明，但不打算强迫他。通过观察现实情况，她解释了用什么方式可以通往更加清晰的未来。她遇到不懂的事情会立刻提问，她能够区分事情的重要性。她清晰地解释了一种快速适应环境的方法。她认为在变色龙身上发生的事情也会发生在她自己身上，最后都会成为生命循环的一部分。她确信那些遭遇痛苦的个体最后会变得比危机来临前更加强大。

变色龙

变色龙还没有做好搬家以及放弃最喜爱的食物的准备。他拒绝改变，并为自己不想适应新环境寻找借口。受到反驳后他让步了，也同意了对方的观点，但他又快速地提出了另一个海藻的建议行不通的理由：种群数量过度增长。不过，面对海藻提出的新的反对意见，他缓和了自己的态度并改变了看法，认为自己能够占据合理的生态位。他将那些对整个生态系统造成影响的变化建立了联系。他不再沮丧，认为自己的家族可以顺利适应变化，这也会让他们能够改变环境。

艺术
The Arts

进化往往是不可预知的。我们来用这个事实探索抽象艺术。看一看维克多·瓦萨雷利的画作《变色龙》。它仅仅包含了圆形、环形、正方形和三角形，且各形状之间留有空白。现在研究一下变色龙的图片，然后只用圆形、环形、正方形和三角形来创作一幅抽象画作。不是每个人都能理解你的作品或者明白你画的是什么，但是重要的是，表达出你是如何想象变色龙的进化历程的。

思维拓展
Systems: Making the Connections

万事万物都在变化，哲学家赫拉克利特说："变化是生命中唯一不变的。"当下的时代被称为人类世，这意味着人类是改变周遭世界的主要力量。在自然界中，物种很快就适应了，科学证明有许多变化正在发生——比我们想象的时间范围要短得多。然而并非所有的改变都是永久性的。有些没有被物种的后代们继承。然而，一旦发生变化，我们确实能看到一系列由变化带来的适应过程。这意味着一个有机体的变化会引发其他有机体的变化。这种转变由环境引起，又会反过来作用于环境。人类似乎对自己行为产生的多米诺骨牌效应并没有任何了解。比如，一种更大的蜥蜴被引入变色龙所居住的岛屿，变色龙就不得不去适应在灌木上生活。此后，以前在地面生活时所捕食的蜘蛛则会摧毁岛上的昆虫种群，导致地面生态发生变化。这将使整个环境变得愈发脆弱，遭遇越来越多无法控制的变化。如果再不做出改变，将错失可持续发展的未来。

动手能力
Capacity to Implement

我们如何激励他人改变？如果我们希望事情有所变化，我们确实需要激励其他人加入进来，以适应新的情境。即使人们在理性上同意改变，为了让他们意识到自己的习惯是如何妨碍自己进行改变，可以和他们一起做以下练习：让一群人交叉双臂。检查哪个手腕在上面，左边还是右边？现在让他们放松并伸展手臂。接下来，让他们在你数到三的时候再次交叉双臂，但这次是让另一只手腕放在上面。大多数人一开始不会成功，或者当被要求再做几次的时候，他们会慢慢地回到最适合自己的交叉手臂的方式。一旦他们意识到自己在这样做，他们就更容易进行一些改变。

故事灵感来自
This Fable Is Inspired by

丽贝卡 · J · 贝斯特
Rebecca J. Best

丽贝卡·J·贝斯特于 2003 年获得不列颠哥伦比亚大学环境科学学士学位。2007 年，她在同一所大学获得森林科学硕士学位。她于 2008 年在加州大学戴维斯分校获得人口生物学博士学位。贝斯特博士在瑞士联邦水产科学与技术研究所（EAWAG）做了两年博士后研究。她是一名水生群落和进化生态学家，致力于研究是什么过程决定了一个特定群落中共存的物种的数量和类型。贝斯特博士目前是美国北亚利桑那大学地球与可持续发展学院的副教授。在这里，她研究温度和降水的变化对水生连通性、物种分布和干旱景观遗传多样性的影响。

图书在版编目(CIP)数据

冈特生态童书.第七辑:全36册:汉英对照 /
(比)冈特·鲍利著;(哥伦)凯瑟琳娜·巴赫绘;
何家振等译.—上海:上海远东出版社,2020
ISBN 978-7-5476-1671-0

Ⅰ.①冈… Ⅱ.①冈… ②凯… ③何… Ⅲ.①生态
环境–环境保护–儿童读物—汉英 Ⅳ.①X171.1-49

中国版本图书馆CIP数据核字(2020)第236911号

策　　划	张　蓉
责任编辑	程云琦
助理编辑	刘思敏
封面设计	魏　来　李　廉

冈特生态童书

你能变得多快?

[比]冈特·鲍利　著
[哥伦]凯瑟琳娜·巴赫　绘

贾龙慧子　译

记得要和身边的小朋友分享环保知识哦!
八喜冰淇淋祝你成为环保小使者!